蔬菜就該這樣炒

目錄 Contents

4 西蘭花 (綠花椰菜)• 白花椰菜主要保健功效／營養烹調祕訣
Broccoli • Cauliflower Health Benefits/Nutritional Facts and Cooking Tips

5 甜豆西蘭花
Stir-fried Sweet Peas and Broccoli

6 西蘭花炒雙鮮
Stir-fried Broccoli with Seafood
怎麼看花椰菜是不是新鮮？
How to check the freshness of broccoli and cauliflower?

8 莧菜主要保健功效／營養烹調祕訣
Chinese Amaranth (Yin Choy) Health Benefits/Nutritional Facts and Cooking Tips

吻仔魚莧菜
Yin Choy with White Baits

9 金銀蛋莧菜
Yin Choy with Century Egg and Salted Egg

10 菠菜主要保健功效／營養烹調祕訣
Spinach Health Benefits/Nutritional Facts and Cooking Tips

11 菠菜丸子湯
Spinach and Prawn Balls Soup

12 炒木須肉
Muxu Pork
炒菜好吃——鍋氣很重要！
Tips to a good stir-fry dish.

14 芥蘭主要保健功效／營養烹調祕訣
Kailan Health Benefits/Nutritional Facts and Cooking Tips

15 芥蘭炒肉
Stir-fried Pork with Kailan

16 空心菜• 菜心 • 青江菜主要保健功效／營養烹調祕訣
Kangkong • Choy Sum•Shanghai Green Health Benefits/Nutritional Facts and Cooking Tips

17 番薯葉 • 油麥菜 • 芥菜主要保健功效／營養烹調祕訣
Sweet Potato Leaves • Leaf Lettuce • Mustard Green Health Benefits/Nutritional Facts and Cooking Tips

18 巴拉煎炒空心菜
Kangkong Belacan

19 青菜拌肉燥
Green Leafy Vegetables with Braised Minced Pork

20 大白菜 • 高麗菜主要保健功效／營養烹調祕訣
Chinese Cabbage • Cabbage Health Benefits/Nutritional Facts and Cooking Tips

21 蝦米炒大白菜
Chinese Cabbage with Dried Shrimps

22 香菇大白菜燒麵筋
Braised Chinese Cabbage with Mushrooms and Wheat Gluten Balls
大白菜先洗後切，防止營養流失！
Wash Chinese cabbage before cutting to prevent nutrients loss.

24 咖哩雜菜
Mixed Vegetables Curry
高麗菜湯汁含豐富維生素C，千萬別浪費！
Drink up the goodness of nutrients in cabbage soups and gravies.

26 芋頭 • 蓮藕 • 山藥主要保健功效／營養烹調祕訣
Yam • Lotus Root • Chinese Yam Health Benefits/Nutritional Facts and Cooking Tips

27 蓮藕章魚豬肉湯
Lotus Root Soup with Dried Octopus and Pork

28 香芋魚頭煲
Claypot Fish Head with Yam
燜煮芋頭要注意的事！
How do you cook yam so that it does not turn mushy?

30 山藥花菇鳳爪湯
Chinese Yam Soup with Mushrooms and Chicken Feet
處理山藥的基本要領！
What should you take note of when preparing Chinese yam?

32 白蘿蔔主要保健功效／營養烹調祕訣
Radish Health Benefits/Nutritional Facts and Cooking Tips

33 白蘿蔔燜魚膠
Braised Radish and Fish Paste

34 蘆筍主要保健功效／營養烹調祕訣
Asparagus Health Benefits/Nutritional Facts and Cooking Tips

35 蘆筍炒蝦仁
Stir-fried Asparagus and Prawns

36 馬鈴薯主要保健功效／營養烹調祕訣
Potato Health Benefits/Nutritional Facts and Cooking Tips

絞肉馬鈴薯
Potato with Minced Pork

37 馬鈴薯洋蔥湯
Potato Onion Soup

38 黃瓜•冬瓜•絲瓜主要保健功效／營養烹調祕訣
Cucumber • Winter Melon • Loofah Health Benefits/Nutritional Facts and Cooking Tips

39 蛤蜊絲瓜粥
Clams and Loofah Porridge

40 雙耳煮絲瓜
Loofah with Black and White Fungus
絲瓜色澤翠綠的祕訣！
How do you cook loofah so that it stays green and does not turn mushy?

42 釀節瓜
Stuffed Hairly Gourd
還有哪些食材適合用來釀肉？
What other kinds of vegetables may be used for this dish?

44 苦瓜主要保健功效／營養烹調祕訣
Bitter Gourd Health Benefits/Nutritional Facts and Cooking Tips

涼拌五味苦瓜
Cold Bitter Gourd with Five-flavours Sauce

45 江魚仔參巴醬燜苦瓜
Bitter Gourd with Ikan Bilis Sambal

46 南瓜主要保健功效／營養烹調祕訣
Pumpkin Health Benefits/Nutritional Facts and Cooking Tips

47 蝦米南瓜
Pumpkin with Dried Shrimps

48 菇菌主要保健功效／營養烹調祕訣
Fungi Health Benefits/Nutritional Facts and Cooking Tips

49 什錦炒菇菌
Stir-fried Assorted Mushrooms

50 金針菇牛肉捲
Beef Rolls with Enoki Mushrooms
切牛肉有技巧！
How to cut beef for the best texture?

52 四季豆•豌豆•長豆•秋葵主要保健功效／營養烹調祕訣
French Bean • Pea • Long Bean • Lady's Finger Health Benefits/Nutritional Facts and Cooking Tips

53 辣蝦米秋葵
Stir-fried Lady's Finger with Spicy Dried Shrimp

55 四季豆燒粉絲
Braised French Beans with Tung Hoon
泡粉絲要用冷水或溫水？
Do you soak tung hoon in cold or warm water?

57 四季豆魚麵
Noodles with French Beans and Fish
去除粗筋功夫不可省！
Be sure to trim away the tough fibers.

58 茄子•甜椒•番茄主要保健功效／營養烹調祕訣
Eggplant • Capsicum • Tomato Health Benefits/Nutritional Facts and Cooking Tips

59 蔬菜天婦羅
Vegetable Tempura

花椰菜 ——
西蘭花(綠花椰菜)、白花椰菜
Broccoli & Cauliflower

蔬菜沙拉
Vegetable Salad

西蘭花魚片湯
Broccoli Fish Soup

芝士焗西蘭花
Cheese Baked Broccoli

主要保健功效：
Health Benefits:

含大量的礦物質、維生素、葉酸、膳食纖維等。具有排毒、強力抗氧化、增強免疫力、清理血管和加強微血管壁強韌度的功效。有助預防感冒、高血壓、糖尿病、心血管疾病和預防腫瘤等功效。綠花椰菜被譽為十字科植物之王。白花椰菜含少量致甲狀腺腫的物質，缺乏碘症狀的人不宜食用。

A good source of minerals, vitamins, folate and dietary fibre. Helps detoxify, high in anti-oxidants, enhances immunity, cleanses the blood and strengthens capillary walls. Helps prevent colds and flu, high blood pressure, diabetes, cardiovascular disease and tumors. Broccoli is known as the king of the brassica family. Cauliflower contains a natural substance that can interfere with the function of thyroid gland, thus avoid eating large amount of cauliflower if you lack in iodine.

營養烹調秘訣：
Nutritional Facts and Cooking Tips:

花椰菜含的維生素C很容易在空氣中被氧化而造成營養流失，所以不宜久收，最好盡快食用。由於花椰菜有澀汁，所以一定要在水中煮一下再食用。花椰菜烹調前，先過熱水但不宜過久，再用少許的油，大火快炒，這樣就可以留住更多的營養成分。

Broccoli and cauliflower loses its vitamin C easily, thus it is best to consume it as soon as possible. Broccoli and cauliflower tastes tart and bitter when eaten raw, cook them in water before eating. Blanch briefly in water, then stir-fry in a little oil to retain most of its flavour and nutrients.

花椰菜也可以這樣吃！
Serving Suggestions:

燙熟後可做成沙拉或涼拌，也可加其他配料炒燴或煮湯。熟悉的菜餚有清炒西蘭花、西蘭花百合蝦仁、芝士焗西蘭花等等。

Use blanched broccoli and cauliflower for salads, or stir-fry with other ingredients. May also be cooked in soups. Familiar dishes include stir-fried broccoli, broccoli with prawns and lily bulbs, cheese baked broccoli.

甜豆西蘭花
Stir-fried Sweet Peas and Broccoli

材料

西蘭花	450克	蘑菇	5朵
甜豆莢	150克	蒜	5瓣
紅蘿蔔	數片	鹽	適量
白果仁	10粒		

做法

1. 西蘭花切小朵，汆燙瀝乾，沖冷水備用；甜豆莢去除頭尾和莢邊粗筋；白果仁去心；蘑菇切片；蒜切末。
2. 用3大匙油爆香蒜末，加入所有材料快炒拌勻，加入少許水以防焦黃，拌炒至熟，加鹽調味，拌炒均勻即可。

Ingredients

450g broccoli	5 button mushrooms
150g sweet peas	5 cloves garlic
some carrot slices	dash of salt
10 ginkgo nuts	

Method

1. Cut broccoli into small florets, blanch and drain well. Rinse under cold water and set aside. Trim sweet peas, cutting the ends off. Remove the cores of the ginkgo nuts. Cut button mushrooms into slices and chop garlic.
2. Fry garlic using 3 tbsps of oil until fragrant, add all the ingredients and toss well briskly, add a little water to prevent burning. When the vegetables are cooked, season with salt and stir well.

怎麼看花椰菜是不是新鮮？
How to check the freshness of broccoli and cauliflower?

新鮮的花椰菜，花蕾密集緊實，除了用眼看，也可用手掌輕輕觸壓感覺，如果花蕾呈枯黃色即是不新鮮了！

Fresh broccoli or cauliflower is at its best when the buds are tight and firm to the touch. It should not yellowish in colour.

西蘭花炒雙鮮
Stir-fried Broccoli with Seafood

Tasty Tips

- 海鮮肉質鮮嫩，非常容易熟，無論是過油、汆燙或炒煮都不宜過久，以免肉質變老。
- *Seafood cooks very fast, thus be it scalding in oil, blanching or stir-frying, it is important not to overcook them in case they turn tough.*

材料

西蘭花	400克
紅蘿蔔	10片
白魚肉	250克
花枝	1只
蔥	1支
薑	8小片
太白粉水	2茶匙

調味料

水或高湯	1/3杯
鹽	1/4茶匙
酒	2大匙
麻油	少許

醃料

鹽	1/4茶匙
胡椒粉	少許
太白粉	1/2大匙
薑汁	1茶匙

做法

1. 魚肉切塊，加入醃料醃製10分鐘以上；蔥切段。
2. 花枝處理乾淨，切紋、切塊，放入70-80°C熱水中汆燙，一轉白即可撈出。
3. 西蘭花切小朵，用滾水燙煮3分鐘，水中加入少許鹽和油，撈起裝盤。
4. 燒熱2杯油，放入魚塊過油至8分熟。
5. 另起油鍋燒熱一匙油，放入蔥段及薑片爆香，加入魚塊、西蘭花、花枝和紅蘿蔔，再加入調味料，輕輕拌炒，加入太白粉水勾芡。

Ingredients

400g broccoli
10 slices carrot
250g white fish fillet
1 cuttlefish
1 stalk spring onion
8 small slices ginger
2 tsps potato starch mixture

Seasonings

⅓ cup water or stock
¼ tsp salt
½ tbsp wine
dash of sesame oil

Marinade

¼ tsp salt
dash of pepper
½ tbsp potato starch
1 tsp ginger juice

Method

1. Cut fish into pieces and mix well with marinade. Set aside for at least 10 minutes. Cut spring onion into sections.
2. Clean cuttlefish and score on the surface before cutting into pieces. Scald in 70-80°C hot water. Remove when it turns white.
3. Cut broccoli into small florets. Add a dash of salt and oil to the boiling water and blanch broccoli in the water for 3 minutes.
4. Heat 2 cups of oil and briefly scald fish in the oil until 80% cooked.
5. In a separate pan, heat 1 tbsp of oil, fry spring onion and ginger until fragrant. Add fish, broccoli, cuttlefish, carrot and Seasonings. Toss well gently, then thicken with potato starch mixture.

莧菜 Chinese Amaranth

主要保健功效：
Health Benefits:

具有補血、增強造血功能和強壯骨骼和牙齒的重要元素。莧菜的鈣質含量比鮮奶高，鐵質含量比菠菜高，又不像菠菜含有草酸，容易被人體吸收。但莧菜易造成小產，婦女懷孕前期避免食用。

Nourishes blood, aids in formation of blood cells and strengthens the bones and teeth. Chinese amaranth contains more calcium than milk and has a higher iron content than spinach. It does not have oxalic acid (which spinach does) and is easily digested by the human body. However it is not suitable for pregnant woman in the early trimesters.

營養烹調祕訣：
Nutritional Facts and Cooking Tips:

莧菜含豐富的鐵和鈣，搭配對的食材，可以增加營養的吸收率！莧菜配上豐富蛋白質的食材可助鐵質的吸收。與動物性蛋白或植物性高鈣食品，如蝦、豆腐、吻仔魚和蛋等同食，能促進鈣質吸收。

It is rich in iron and calcium and when paired with the right foods, it helps in the absorption of nutrients. Cook with protein rich ingredients to aid in the absorption of iron, and animal protein rich or calcium rich plant-based foods to help in the absorption of calcium, these include prawns, tofu, white baits and eggs.

吻仔魚莧菜
Yin Choy with White Baits

材料
莧菜 300克
(去老筋，摘成段)
吻仔魚 1/2杯
蒜末 1茶匙
蔥段 少許

調味料
酒 1茶匙
鹽 1/3茶匙
太白粉水 少許

做法
1. 鍋中放1大匙油燒熱，爆香蔥段，加入吻仔魚炒片刻，淋酒，盛出。
2. 另起油鍋燒熱一匙油，爆香蒜末，加入莧菜，以大火拌炒片刻，淋下水2/3杯，加鹽調味，煮至莧菜軟，放入吻仔魚略拌，加入太白粉水勾芡即可。

Ingredients
300g yin choy
(remove stringy parts, pluck into sections)
½ cup white baits
1 tsp chopped garlic
some spring onion sections

Seasonings
1 tsp wine
⅓ tsp salt
some potato starch mixture

Method
1. Heat 1 tbsp of oil, fry spring onion until fragrant. Toss in white baits and drizzle over wine. Remove.
2. Heat 1 tbsp of oil, fry garlic until fragrant. Stir in yin choy and fry for a while over high heat. Pour over ⅔ cup of water. Season with salt and cook until yin choy is soft. Toss in white baits and thicken with potato starch mixture.

金銀蛋莧菜

Yin Choy with Century Egg and Salted Egg

材料		調味料	
莧菜	500克	鹽	適量
皮蛋	2個	胡椒粉	適量
鹹蛋	1個	酒	1茶匙
蒜末	1茶匙	麻油	1茶匙
薑片	少許		
高湯或水	2杯		

做法

1. 莧菜去除老筋，摘成段；鹹蛋和皮蛋蒸熟，剝殼，切丁。
2. 鍋中放少許油燒熱，爆香蒜和薑，加入莧菜、高湯或水，煮至滾、莧菜微軟，加入鹹蛋、皮蛋和調味料煮片刻即可。

Ingredients

500g Chinese amaranth (yin choy)
2 century eggs
1 salted egg
1 tsp minced garlic
some ginger slices
2 cups stock or water

Seasonings

dash of salt
dash of pepper
1 tsp wine
1 tsp sesame oil

Method

1. Remove the stringy parts from the yin choy and pluck into sections. Steam salted egg and century eggs until cooked, shell and cut into dices.
2. Heat oil and fry garlic and ginger until fragrant. Add yin choy, stock or water and bring to a boil. When the yin choy is fairly soft, add salted egg, century eggs and Seasonings. Cook briefly before serving.

菠菜
Spinach

菠菜豬肝湯
Spinach and Pig's Liver Soup

柴魚菠菜
Spinach and Bonito

菠菜炒臘肉
Stir-fried Spinach with Cured Meat

主要保健功效：
Health Benefits:

含有豐富的維生素A、B群、C、胡蘿蔔素、葉酸、膳食纖維、鐵、鈣、鉀等。具有保護眼睛、有益胎兒大腦神經發育、補血、抗衰老、預防骨骼疏鬆症、通便等功效。

Rich in vitamin A, C, B-group vitamins, beta-carotene, folate, dietary fibre, iron, calcium and potassium. Packed full of nutrition, spinach protects the eyes, aids in babies' brain development, nourishes blood, anti-aging, prevents osteoporosis and promotes bowel movements.

營養烹調祕訣：
Nutritional Facts and Cooking Tips:

菠菜所含的「草酸」成分容易和鈣結合成結石。菠菜汆燙或炒，可煮熟一些，草酸受高溫就會被破壞，去除菠菜中過多的草酸，有助鈣和鐵的攝取，同時也可以去除其苦澀味。用油炒菠菜，有助胡蘿蔔素的溶解，加快人體的吸收。

The oxalate content in spinach gets combined with calcium and is the cause of kidney stones. Blanching or stir-frying spinach lowers the oxalate content to aid the absorption of calcium and iron, and also removes its tartness. Stir-frying it in oil makes it easier for the body to absorb the beta-carotene it contains.

菠菜還可以這樣吃！
Serving Suggestions:

菠菜適合炒、汆燙後涼拌、煮成羹。熟悉的菜餚有菠菜豬肝湯、菠菜羹等等。

Suitable for stir-frying and blanching before using in salads or cooked in soups. Common dishes include spinach and pig's liver soup, spinach thick soup.

菠菜丸子湯
Spinach and Prawn Balls Soup

材料

菠菜	120克
豆腐	1盒
蝦丸或魚丸	8粒
竹笙	2條

調味料

鹽	適量
雞粉	適量

做法

1. 菠菜切段，略汆燙；竹笙泡發，切段；豆腐切片。
2. 鍋中放入4杯水燒開，放入丸子，蓋上鍋蓋，煮至丸子浮起，加入竹笙和調味料煮沸，加入豆腐和菠菜煮熟即可。

Ingredients

120g spinach
1 box tofu
8 prawn balls or fish balls
2 strips bamboo fungus

Seasonings

dash of salt
dash of chicken seasoning powder

Method

1. Cut spinach into sections and briefly blanch in water. Rehydrate bamboo fungus and cut into sections. Cut tofu into slices.
2. Bring 4 cups of water to a boil. Place prawn or fish balls into the water, put the lid on and cook until they float to the surface. Add bamboo fungus and Seasonings. Bring to a boil, then add spinach and tofu to cook through.

炒菜好吃——鍋氣很重要！

Tips to a good stir-fry dish.

炒菜時沿著鍋邊淋一點水，可使鍋中產生水氣、鍋氣，帶著水氣，容易使各種食材和調味料滋潤、融合，使食材產生香氣而變得更好吃。

During stir-frying, drizzle some water along the sides of the wok, this will create steam and “wok hei”, the “essence” imparted by a hot wok which allow the flavours of the ingredients and seasonings to come together and give a good stir-fry dish its distinctive flavour and aroma.

炒木須肉
Muxu Pork

Tasty Tips

- 這道菜因為把蛋炒成小塊、細碎狀，好像木須花（桂花）的形狀而得名。
- 可配薄餅上桌包食。
- The scrambled eggs in this dish resemble a certain kind of flower by the name of Muxu, thus the Chinese name Muxu Pork.
- May be served wrapped in crepes.

材料

菠菜	120克
肉絲	120克
水發木耳	1杯
筍	1支
蛋	2個
蔥花	1大匙

調味料A

醬油	1/2大匙
太白粉	1/2大匙
水	1大匙

調味料B

醬油	1大匙
鹽	1/4茶匙

做法

1. 肉絲用調味料A拌勻，醃制約10分鐘，下鍋前加入1大匙油拌勻。
2. 菠菜、木耳、筍洗淨。菠菜切段；木耳切絲；筍煮熟後切絲。
3. 燒熱2大匙油至160°C，放入肉絲，炒至肉絲變色後盛出，瀝乾油。
4. 蛋加入1/4茶匙鹽打散後，用少許油炒熟，盛出。
5. 另起油鍋燒熱少許油，先爆香蔥花，放入筍絲、木耳絲和少許水炒勻，加入調味料B，再加入菠菜炒熟，最後加入肉絲和蛋，翻炒均勻即可。

Ingredients

120g spinach
120g pork slices
1 cup rehydrated black fungus
1 stalk bamboo shoot
2 eggs
1 tbsp chopped spring onion

Seasonings A

½ tbsp dark soy sauce
½ tbsp potato starch mixture
1 tbsp water

Seasonings B

1 tbsp dark soy sauce
¼ tsp salt

Method

1. Marinate pork slices with Seasonings A and set aside for about 10 minutes. Stir in a tbsp of oil before cooking.
2. Rinse vegetables. Cut spinach into sections and black fungus into shreds. Cook bamboo shoot and cut into shreds.
3. Heat 2 tbsps of oil to 160ºC, add pork and fry until it changes colour. Remove and drain away the oil.
4. Beat eggs with ¼ tsp of salt. Scramble the eggs in a little oil and dish out.
5. Heat oil and fry chopped spring onion until fragrant, stir in bamboo shoot, black fungus and a little water. Add Seasonings B, then toss in spinach to cook through. Lastly add pork and eggs to combine well.

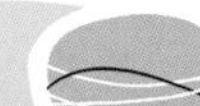

芥蘭
Kailan

炒什錦蔬菜
Stir-fried Vegetables

清炒芥蘭
Stir-fried Kailan

玉蘭雞
Chicken and Ham with Kailan

主要保健功效:
Health Benefits:

含豐富的維生素B群、C、蛋白質、膳食纖維、鈣、磷、鐵等。芥蘭是低草酸高鈣的蔬菜,有助防止骨質疏鬆、幫助睡眠、保健視力、增強免疫力、防癌等。

Contains B-group vitamins, vitamin C, protein, dietary fibre, calcium, phosphorus and calcium. Low in oxalic acid but high in calcium, kailan helps prevent osteoporosis, aids sleep, promotes good vision, enhances immunity and is anti-cancer.

營養烹調祕訣:
Nutritional Facts and Cooking Tips:

如同其他綠葉蔬菜,芥蘭含水溶性維生素,因此清洗蔬菜時不宜浸泡水中太久,以免流失營養素。芥蘭不適合長時間烹煮,所以最好是用來快炒或汆燙,這樣才能保持色澤和口感清脆。芥蘭帶有苦澀味,烹調時加入薑汁有助去除苦澀味。也可加些糖或酒,具有去除苦澀味和增香的作用。

As with other green vegetables, kailan contains water-soluble vitamins, thus it is advised not to soak it in water for too long when washing to prevent loss of nutrients. Briskly stir-fry or blanch kailan to best preserve its colour and crispness. Adding ginger juice helps get rid of its bitterness. Sugar and wine work the same way too and also add fragrance to the dish.

芥蘭還可以這樣吃!
Serving Suggestions:

適合汆燙後清炒或加入配料一起炒。常見的菜餚有廣式鹹魚炒芥蘭、蠔油芥蘭;潮州人的大地魚炒芥蘭、芥蘭排骨湯和一般家常的芥蘭炒牛肉、芥蘭炒魚片等等。

Suitable for blanching then stir-frying by itself or with other ingredients. Common dishes include Cantonese style stir-fried kailan with salted fish and kailan in oyster sauce, Teochew style stir-fried fish with kailan and kailan pork ribs soup. Other family favourites include stir-fried beef or fish with kailan.

芥蘭炒肉
Stir-fried Pork with Kailan

材料

芥蘭菜	200克
肉片	100克
蔥段	少許

調味料

醬油	1茶匙
鹽、酒、麻油	各少許

做法

1. 肉片加入少許醬油、水和太白粉拌勻，醃漬約10分鐘。
2. 芥蘭菜切段。先放入菜梗炒透，再加入菜葉，加少許酒、鹽和糖調味，盛盤。
3. 鍋中放2大匙油燒熱，爆香蔥段，放入肉片炒至8分熟，再放入芥蘭菜拌炒片刻，淋下醬油、1大匙水和鹽炒勻，最後滴入酒和麻油即可。

Ingredients

200g kailan
100g pork slices
some spring onion sections

Seasonings

1 tsp dark soy sauce
dash of salt, wine, sesame oil

Method

1. Marinate pork slices with a little dark soy sauce, water and potato starch, set aside for about 10 minutes.
2. Cut kalian into sections. Fry the stem first, then the leaves. Season with wine, salt and sugar. Remove.
3. Heat 2 tbsps of oil and fry spring onion until fragrant, add pork and fry until 80% cooked. Stir in kailan and fry briefly. Add dark soy sauce, 1 tbsp of water and a dash of salt. Combine well and drizzle in wine and sesame oil before serving.

空心菜 Kangkong

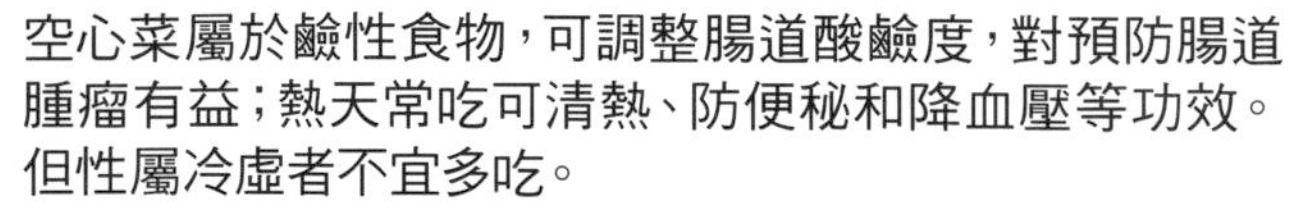

主要保健功效

Health Benefits:

空心菜屬於鹼性食物，可調整腸道酸鹼度，對預防腸道腫瘤有益；熱天常吃可清熱、防便秘和降血壓等功效。但性屬冷虛者不宜多吃。

Kangkong is an alkaline food vital for helping to maintain intestinal pH balance, thus preventing intestinal tumors. When eaten on a hot day, it helps dispel heat and cool the blood, prevents constipation and lowers blood pressure. It is cold in nature and is not suitable for people of a weak body constitution.

營養烹調祕訣

Nutritional Facts and Cooking Tips：

空心菜中所含葉綠素有「綠色精靈」之稱；空心菜遇熱容易變黃，烹調時要充分熱鍋，大火快炒，不等葉子變軟即可熄火。

Kangkong has a high chlorophyl level. It turns yellow easily when heated. Make sure you heat up the wok and briskly fry it over high heat, dish out before the leaves turn soft.

菜心 Choy Sum

主要保健功效

Health Benefits:

具有促進成長發育、減輕骨骼酥鬆症、清熱解毒、散血消腫等功效。菜心花亦有豐富的維生素C、K、B2和胡蘿蔔素，很適合口腔易潰爛、牙齦易出血、皮膚乾燥易癢的人食用。

Promotes growth, eases osteoporosis, dispels heat and detoxifies the body, reduces the risk of blood clots and relieves swelling. Choy sum flowers also contain vitamins C, K, B2 and beta-carotene. It is suitable for people prone to mouth ulcers, bleeding gums and dry and itchy skin.

營養烹調祕訣

Nutritional Facts and Cooking Tips：

菜心又叫油菜，適合用來快炒，不適合長時間加熱。吃剩的菜心隔夜後不宜食用，以免吃下沈澱致癌物質——亞硝酸鹽，得不償失！

Suitable for quick stir-fries. Not suitable for prolonged cooking. Do not eat cooked choy sum left overnight as it contains nitrite which causes cancer.

青江菜 Shanghaí Green

主要保健功效

Health Benefits:

具有養眼美容，改善便秘、清除內熱、滋潤皮膚、防止老化等功效，對於保護眼睛和牙齒也有很大的幫助。

Benefits the eye, complexion and teeth. It also helps prevent constipation, dispels body heat, moisturizes the skin and has anti-aging properties.

營養烹調祕訣

Nutritional Facts and Cooking Tips：

適合快炒，即火候大，油熱後下辛香料熗炒出香味，再下菜，並快速起鍋。加熱的時間不宜過長，以免破壞營養成分，也影響口感。

Suitable for quick stir-fries. Cook in the shortest time possible to retain its nutrients and texture.

番薯葉 Sweet Potato Leaves

主要保健功效

Health Benefits:

番薯葉含豐富的膳食纖維，可幫助排出體內油脂，減低體內壞膽固醇，促進腸胃蠕動；其他營養成分還有助抗氧化、分解毒素、增強免疫力等。

Contains dietary fibre which helps dispel fats from the body, lowers bad cholesterol level and promotes bowel movements. Also contains other nutrients with anti-oxidant benefits, breaks down toxic and enhances immunity.

營養烹調祕訣

Nutritional Facts and Cooking Tips：

昔日的飼豬食材，如今被列為十大抗氧化蔬菜之一。番薯葉適合快炒，不可生食，一定要煮熟後才食用。用油爆炒，並連同湯汁一起吃，最能充分攝取到營養成分。

Sweet potato leaves are now known as one of the top ten anti-oxidant vegetables. They are suitable for quick stir-frying and not suitable for eating raw. Briskly fry over high heat in oil, then serve with the gravy to get the full benefits of the vegetable.

油麥菜 Leaf Lettuce

主要保健功效

Health Benefits:

具有清肝、利尿、消除緊張、幫助睡眠、降脂等功效。油麥菜性寒涼，同時對視神經有刺激作用，體質虛寒者和有眼疾者不宜多吃。

Helps cleanse the liver, promotes urination, relieves tension, promotes sleep and lowers fats. Leaf lettuce is cold in nature and might have an effect on vision, thus people with cold and weak constitution or with eye problems should refrain from eating leaf lettuce.

營養烹調祕訣

Nutritional Facts and Cooking Tips：

炒的時間不可太久，去生味即可食用，以免破壞其營養成分，同時失去脆嫩的口感和鮮綠色澤。油麥菜非常適合淋醬生吃！

Stir-fry briefly so that it stays tender crisp and green. Delicious eaten raw with a dressing.

芥菜 Mustard Green

主要保健功效

Health Benefits:

一種很好的鹼性蔬菜，具有促進腸胃蠕動和食慾、幫助新陳代謝等功效。

An alkaline vegetable good for maintaining intestinal health, increasing appetite and promoting metabolism.

營養烹調祕訣

Nutritional Facts and Cooking Tips：

常被製成醃製品，但鹽分極高，容易對腎臟造成負擔，最好是鮮食。

It is commonly used to make preserved vegetables. However the salt content of preserved vegetables is very high and might put a strain on your kidneys, thus it is best not to consume too much of these.

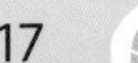

巴拉煎炒空心菜
Kangkong Belacan

材料

空心菜	400克（切段）
巴拉煎	1 1/2大匙（煸香）
蝦米	2大匙（泡軟）
辣椒	2支（切片）
蒜	3瓣（切末）

調味料

醬油	1茶匙
糖	1茶匙
雞粉	1/2茶匙

做法

1. 燒熱2大匙油，放入蒜茸和蝦米爆香，加入巴拉煎拌炒至有香味，加入空心菜和少許水，以大火快炒均勻。
2. 加入辣椒和調味料拌炒，至空心菜變軟即可。

註：巴拉煎為蝦醬的一種。

Ingredients
400g kangkong (cut into sections)
1½ tbsps belacan (toasted)
2 tbsps dried shrimps (soak till soft)
2 chillies (sliced)
3 cloves garlic (chopped)

Seasonings
1 tsp dark soy sauce
1 tsp sugar
½ tsp chicken seasoning powder

Method
1. Heat 2 tbsps of oil, fry garlic and dried shrimps until fragrant. Add belacan and fry until fragrant. Stir in kangkong and a little water, combine well briskly over high heat.
2. Lastly add chillies and Seasonings to fry until kangkong is soft.

青菜拌肉燥
Green Leafy Vegetables with Braised Minced Pork

材料
青菜　　250克
肉燥醬　1-2大匙

調味料
鹽　　1茶匙

做法
1. 青菜掰開洗淨；肉燥醬放入微波爐加熱。
2. 鍋中放4杯水燒開，加入鹽，再放入青菜燙熟，撈出瀝乾水，排盤，淋上肉燥醬即可。

Ingredients
250g green leafy vegetables
1-2 tbsps braised minced pork sauce

Seasonings
1 tsp salt

Method
1. Rinse and trim the vegetables. Heat up the braised minced pork in the microwave oven.
2. Bring 4 cups of water to a boil, add salt and blanch vegetables till cooked. Drain well and place on a plate. Pour over braised minced pork sauce.

大白菜 Chinese Cabbage

主要保健功效
Health Benefits:

具有清熱退火、幫助消化、抗氧化、排毒利尿、預防便秘、心血管疾病和感冒等功效。

Effective in helping to dispel heat from the body, aids digestion, high in anti-oxidants, detoxify the body, has a diuretic effect and prevents constipation, cardiovascular disease, colds and flu.

營養烹調祕訣
Nutritional Facts and Cooking Tips：

大白菜含大量水溶性維生素，所以不宜長時間泡水或燙，以免維生素大量流失在水中。適合炒燴，尤其適合煮湯，可以從湯汁中攝取到營養素，且更容易被消化吸收。

Chinese cabbage contains water-soluble vitamins that are easily lost during prolonged soaking and boiling. Suitable for stir-frying, especially for cooking in soups as we can easily digest the nutrients in the soup.

大白菜也可以這樣吃！
Serving Suggestions:

適合炒、燴、燉煮、滾湯，也可以做成泡菜、菜乾。熟悉的菜餚有大白菜滾湯、蝦米炒大白菜、蠔乾發菜燜大白菜等等。

Suitable for stir-frying, braising, stewing and cooking in soups. Also good for making kim chi and preserved vegetables. Familiar favourites include Chinese cabbage soup, Chinese cabbage with dried shrimps, braised Chinese cabbage with dried oysters and black hair moss.

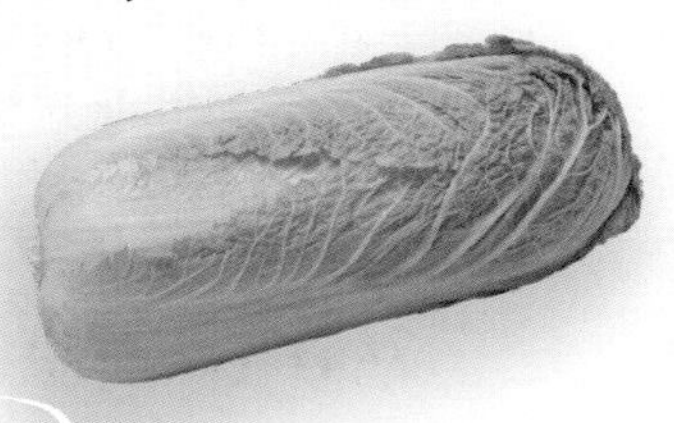

高麗菜 Cabbage

主要保健功效
Health Benefits:

含豐富維生素B群、C、K、U、鈣、磷、鉀、有機酸等。其維生素K具有凝固血液功效，維生素U可以促進胃的新陳代謝和胃粘膜的修復，因此也有「廚房的天然胃菜」的美譽。不過，甲狀腺功能失調者不宜大量食用。

A good source of B-group vitamins, vitamins C, K, U, calcium, phosphorus, potassium and organic acid. Vitamin K is used in the body to control blood clotting, whereas vitamin U helps in regulating gastric disorders and is effective in the healing of gastric mucosa. However it is not suitable for patients with thyroid hormone imbalance.

營養烹調祕訣
Nutritional Facts and Cooking Tips：

高麗菜中所含營養成分不耐高熱！而且其維生素C是屬水溶性，要完全攝取最好生吃或連湯一起食用。

The nutrients in cabbage are easily lost during cooking and it also contains vitamin C which is water-soluble. It is best to eat it raw or drink up the soup that it is cooked in.

高麗菜也可以這樣吃！
Serving Suggestions:

適合炒、燴、煮、滾湯或生吃。熟悉的菜餚有高麗菜滾湯、蝦米炒高麗菜、咖哩雜菜和常吃的蔬菜沙拉等等。

Suitable for stir-frying, braising, boiling, cooking in soups or eaten raw. Familiar dishes include cabbage soup, cabbage with dried shrimps, vegetables curry and coleslaw.

蝦米炒大白菜
Chinese Cabbage with Dried Shrimps

材料

大白菜／絲瓜	400克
蝦米	2大匙
酒	1茶匙
水或高湯	1/2杯
鹽	1/3茶匙

做法

1. 大白菜切寬條；蝦米泡軟，摘去硬殼。
2. 鍋中放1大匙油燒熱，放入蝦米炒香，淋下酒，加入大白菜炒至略變軟，加入水或高湯，可蓋上鍋蓋，將大白菜煮至喜愛的軟熟度，加入鹽調味即可。

Ingredients

400g Chinese cabbage/loofah
2 tbsps dried shrimps
1 tsp wine
½ cup water or stock
⅓ tsp salt

Method

1. Cut Chinese cabbage into wide strips. Soak dried shrimps until soft, then pluck away the hard bits.
2. Heat 1 tbsp of oil and fry dried shrimps until fragrant. Drizzle over wine and add Chinese cabbage to cook until slightly soft. Add water or stock and cover with a lid to cook the Chinese cabbage to your preferred softness. Lastly season with salt.

大白菜先洗後切，防止營養流失！

Wash Chinese cabbage before cutting to prevent nutrients loss.

大白菜含豐富的維生素C，有美容養顏、清熱退火等功效。它的維生素C屬水溶性，處理時先將大片葉子剝下，清洗乾淨再切絲，千萬不要切絲後泡水，以防止維生素C流失。

Chinese cabbage is a good source of vitamin C. It has properties for beautifying the complexion and removing heatiness in the body. As it contains water-soluble vitamins, it is best to tear off the leaves and rinse clean before cutting into shreds. Do not soak the shreds in water, this is to prevent a loss of vitamins.

香菇大白菜燒麵筋
Braised Chinese Cabbage with Mushrooms and Wheat Gluten Balls

Tasty Tips

- 大白菜以葉片包覆緊密結實、無斑點和腐壞的為佳。
- 大白菜宜切絲，較易入味。
- *Choose Chinese cabbage that is firm and dense with leaves free of bruises and blemishes.*
- *It is best to cut Chinese cabbage into shreds as these absorb flavours more readily.*

材料

大白菜	600克
紅蘿蔔	1/2支
香菇	4朵
油麵筋	1杯
蝦米	1大匙
蔥	1支
香菜	少許

調味料

薄鹽醬油	1大匙
鹽	適量
麻油	少許

做法

1. 大白菜洗淨瀝乾水，梗切寬條，葉子切大片；紅蘿蔔切片；蔥切段。
2. 香菇用冷水泡軟，去蒂切片，泡香菇水留用；蝦米泡軟，摘去硬殼；油麵筋用溫水泡軟後略擠乾水。
3. 鍋中放2大匙油燒熱，先爆香蔥段、香菇和蝦米，再淋下醬油，炒香後加入大白菜炒至軟。
4. 加入泡香菇的水和紅蘿蔔，煮約3-5分鐘。
5. 加入油麵筋拌炒均勻，加鹽調味，煮至大白菜已夠軟、湯汁略收乾，滴下麻油、撒下香菜拌勻即可。

Ingredients

600g Chinese cabbage
½ carrot
4 Chinese mushrooms
1 cup fried wheat gluten balls
1 tbsp dried shrimps
1 stalk spring onion
some coriander

Seasonings

1 tbsp light soy sauce
dash of salt
dash of sesame oil

Method

1. Rinse Chinese cabbage. Cut the stalks into wide strips and leaves into big pieces. Cut carrot into slices and spring onion into sections.
2. Soak Chinese mushrooms until soft. Remove the stems and cut into slices. Reserve the water used for soaking the mushrooms. Soak dried shrimps until soft, then pluck away the hard bits. Soak fried wheat gluten balls in warm water until soft, then gently squeeze dry.
3. Heat 2 tbsps of oil and fry spring onion, Chinese mushrooms and dried shrimps until fragrant. Drizzle over light soy sauce, then toss in Chinese cabbage to cook until it turns soft.
4. Add the mushrooms soaking water and carrot to cook for about 3-5 minutes.
5. Add fried wheat gluten balls and stir well. Season with salt. Cook until the Chinese cabbage is soft and the gravy has slightly reduced. Add sesame oil and scatter over coriander to serve.

高麗菜湯汁含豐富維生素C，千萬別浪費！

Drink up the goodness of nutrients in cabbage soups and gravies.

高麗菜的品種很多，常見的有淡綠色的、紫色的，也有小形高麗菜嬰，都含有豐富的水溶性維生素C，比較不適合用來氽燙，較適合用來煮湯，因為維生素C會釋放在水中，連菜帶湯汁一起吃，千萬別浪費！

Cabbage comes in different varieties, light green and purple cabbage and brussel sprouts are the more common ones. Cabbage contains water-soluble vitamin C, so it is not advisable to blanch them before cooking. Cook in soups and drink up the nutrients in the soups.

咖哩雜菜
Mixed Vegetables Curry

Tasty Tips

- 材料可隨意變化，加入馬鈴薯、茄子、秋葵、南瓜、白花椰菜等。
- 椰漿不宜煮太久，以免湯汁變稠。榨汁濃椰漿久煮會煮出椰油味，影響菜餚的味道。
- *Potatoes, eggplants, lady's fingers, pumpkin and cauliflower may also be used for this dish.*
- *Do not cook coconut milk for too long as it will curdle. Freshly squeezed thick coconut milk will give out an oily taste if cooked over a long period of time and this will affect the flavour of the dish.*

材料A

高麗菜	500克
紅蘿蔔	1支
長豆	150克
番茄	2個
洋蔥	1個
油豆腐	4-5塊
小紅蔥	2-3粒
薑	2片

材料B

椰漿	600毫升
巴拉煎	1大匙
咖哩粉	2-3大匙

調味料

鹽	適量
糖	適量

做法

1. 高麗菜切大片；紅蘿蔔切滾刀塊；長豆去除頭尾和莢邊粗筋，折成段；番茄切小瓣；洋蔥切絲；油豆腐切半；小紅蔥切片；巴拉煎乾鍋炒香。
2. 燒熱6大匙油，先炒香洋蔥、小紅蔥和薑，加入巴拉煎炒勻，再加入咖哩粉，翻炒至香味和油溢出。
3. 加入所有蔬菜、油豆腐和調味料，以中小火煮至蔬菜軟透，轉小火加入椰漿略煮即可。

Ingredients A

500g cabbage
1 carrot
150g long beans
2 tomatoes
1 onion
4-5 pieces bean curd puffs
2-3 cloves shallots
2 slices ginger

Ingredients B

600ml coconut milk
1 tbsp belacan
2-3 tbsps curry powder

Seasonings

dash of salt
dash of sugar

Method

1. Cut cabbage into large pieces. Roll-cut carrot into pieces. Remove the ends of the long beans and break into sections. Cut tomatoes into small sections. Cut onion into shreds. Cut bean curd puffs into halves and shallots into slices. Toast belacan in a dry pan until fragrant.
2. Heat 6 tbsps of oil and fry onion, shallots and ginger until fragrant, add belacan and toss well. Stir in curry power and fry until fragrant and oil seeps out.
3. Add all the vegetables, bean curd puffs and Seasonings. Cook over medium-low heat until the vegetables are soft. Turn down the heat, add the coconut milk to cook briefly until heated through.

芋頭 Yam

主要保健功效

Health Benefits:

具有穩定血糖、利尿、解毒等功效。芋頭特有黏性食物纖維，有助刺激腸壁，幫助排便。其黏性蛋白質被人體吸收後，會生成免疫球蛋白，有助增強免疫力。

Stabilizes blood glucose levels, promotes urination and detoxifies the body. Yam's dietary fibre helps establish regular bowel movement. After being absorbed into the body, its protein will turn into immunoglobulin to help boost immunity.

營養烹調祕訣

Nutritional Facts and Cooking Tips：

芋頭含有大量的草酸鈣，煮熟後才能被分解，所以要煮熟後才能食用。它還含有豐富的澱粉和蛋白質，有足夠的營養，也容易有飽足感，因此不宜一次吃太多。

Yam contains calcium oxalate which will only break down upon cooking. Yam has an abundant supply of starch and protein and other nutrients, it makes us feel full easily, thus do not eat too much at a time.

蓮藕 Lotus Root

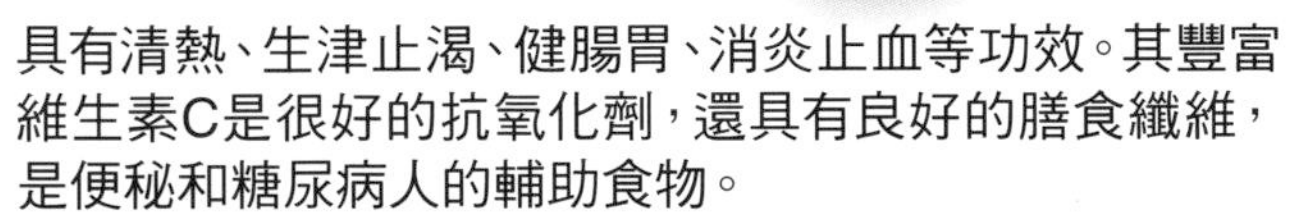

主要保健功效

Health Benefits:

具有清熱、生津止渴、健腸胃、消炎止血等功效。其豐富維生素C是很好的抗氧化劑，還具有良好的膳食纖維，是便秘和糖尿病人的輔助食物。

Dispels heat from the body, promotes the production of bodily fluids, quenches thirst, strengthens the stomach, stops inflammation and bleeding. It is rich in vitamin C which is a powerful anti-oxidant and dietary fibre, which makes it an ideal food for diabetes patients and people suffering from constipation.

營養烹調祕訣

Nutritional Facts and Cooking Tips：

蓮藕可熟吃，也可生吃；連皮生榨成汁，清熱、生津止渴的功效非常顯著，但脾胃虛寒者和消化不佳者，避免生吃，較難消化。

Lotus root may be eaten raw or cooked. Juice the root with the skin on for a drink that is great for dispelling heat, promoting the production of fluids and quenching thirst. However, people with weak stomach or have digestive problems should refrain from eating it raw as it is not easily digested.

山藥 Chinese Yam

主要保健功效

Health Benefits:

其營養成分有助增強免疫力、提高新陳代謝、改善血糖和舒緩女性更年期綜合症狀等功效。其黏液蛋白，有助維持血管彈性，幫助減少脂肪堆積和減低血糖。

Boosts immunity, increases metabolism, improves blood glucose level and relieves menopause symptoms. It contains mucin which helps maintain vascular elasticity, reduces fats accumulation and lowers blood glucose level.

營養烹調祕訣

Nutritional Facts and Cooking Tips：

新鮮的山藥曬乾即是常用的中草藥——淮山。山藥烹調的時間不宜過長，久煮會破壞營養素，造成營養流失。

Fresh Chinese yam can be dried and made into dried Chinese yam (huai shan) which is commonly used as a Chinese medicine. Do not cook Chinese yam too long in order to retain its nutrients.

蓮藕章魚豬肉湯
Lotus Root Soup with Dried Octopus and Pork

材料

蓮藕	450克
豬腱肉	600克
乾章魚	150克
蜜棗	2粒
鹽	適量

做法

1. 蓮藕去節洗淨，切厚片；豬腱肉整塊放入冷水中煮約15分鐘，去除血水，取出後用冷水衝乾淨；乾章魚泡軟，洗淨。

2. 將所有材料放入湯鍋中，加入水2.5公升，以大火煮至滾，轉小火煲約2小時，加鹽調味即可。

Ingredients

450g lotus root
600g pork shank
150g dried octopus
2 honey dates
dash of salt

Method

1. Rinse lotus root and cut into thick slices. Place the whole pork shank in cold water and cook for about 15 minutes, remove and rinse under cold water. Soak dried octopus until soft and rinse well.

2. Place all Ingredients in a soup pot and add 2.5 litres of water. Bring to a boil over high heat, then turn down the heat and simmer for about 2 hours. Season with salt to taste.

燜煮芋頭要注意的事！
How do you cook yam so that it does not turn mushy?

芋頭容易糊化，要注意別燒焦了，尤其是最後湯汁減少時，火要轉小，芋頭夠軟即可熄火。

Yam breaks down and turns mushy easily, thus be sure not to overcook it. Once the gravy or soup has reduced, turn down the heat so that you don't burn the dish. Turn off the heat once the yam has softened.

香芋魚頭煲
Claypot Fish Head with Yam

Tasty Tips

- 芋頭先炸再蒸，比較容易煮軟綿。
- *Deep-fry yam first before steaming so that it softens more easily.*

材料

芋頭	250克
大白菜	150克
草魚頭	2個
蒜苗	35克
芹菜	35克
蝦米	1大匙
薑絲	少許
冬菇	2朵
高湯	4-5杯

調味料

酒	2大匙
麻油	1大匙
鹽	1茶匙
糖	1/2茶匙

做法

1. 魚頭切半；芋頭切滾刀塊；大白菜、蒜苗和芹菜切絲；冬菇泡軟，切絲；蝦米泡軟，去硬殼。
2. 芋頭先炸過，再蒸約8分鐘，取出。
3. 大白菜放入砂鍋中墊底。
4. 魚頭入鍋略煎黃，淋入酒，加入蒜苗、芹菜、蝦米、薑絲和冬菇略炒香，盛入砂鍋。
5. 砂鍋中加入高湯和調味料，煮至滾，加入芋頭，上下拌一下，蓋上砂鍋蓋，燜煮約10分鐘至芋頭夠軟即可。

Ingredients

250g yam
150g Chinese cabbage
2 grass carp fish heads
35g garlic shoot
35g Chinese celery
1 tbsp dried shrimps
some shredded ginger
2 Chinese mushrooms
4-5 cups stock

Seasonings

2 tbsps wine
1 tbsp sesame oil
1 tsp salt
½ tsp sugar

Method

1. Cut fish heads into halves. Roll-cut yam into pieces. Cut Chinese cabbage, garlic shoot and Chinese celery into shreds. Soak Chinese mushrooms until soft and cut into shreds. Soak dried shrimps until soft, then pluck away the hard bits.
2. Deep-fry yam, then steam for about 8 minutes. Remove.
3. Place Chinese cabbage at the bottom of a claypot.
4. Briefly pan-fry fish heads, drizzle over wine, then add garlic shoot, Chinese celery, dried shrimps, ginger and mushrooms. Toss well, then transfer to the claypot.
5. Pour in stock and Seasonings, then bring to a boil. Add yam and stir briefly. Cover with a lid and simmer for about 10 minutes until the yam is soft.

處理山藥的基本要領！

What should you take note of when preparing Chinese yam?

處理山藥最好戴上手套，因為山藥中含有植物鹼，皮膚敏感者可能會引起皮膚發癢。山藥富含鐵質，一接觸空氣會產生氧化變色，所以切塊後要立即放入冰水或鹽水中，就可避免。

People who have skin allergies might get an itch when they handle Chinese yam, thus it is best to put on gloves when preparing it. Chinese yam has a high iron content, it oxidises and turns dark when cut. Soak it in ice water or salted water to prevent discoloration.

山藥花菇鳳爪湯
Chinese Yam Soup with Mushrooms and Chicken Feet

Tasty Tips

- 加入雞胸肉一起煮，湯中會有肉香。
- 花菇燉煮的時間越久越入味好吃，但是它需要長時間的浸泡、處理才能達到最好的效果。
- *Add chicken meat to cook in the soup for added meaty flavour.*
- *Chinese mushrooms taste best when cooked over a long time, for the best result, make sure you rehydrate them properly before cooking.*

材料

山藥	400克
雞爪	10支
雞胸（連骨）	1個
花菇	6朵
紅棗	8-10粒
薑	1塊
太白粉	適量

調味料

酒	1大匙
鹽	適量

做法

1. 山藥去皮，切塊；雞爪和雞胸放入滾水中燙煮約1分鐘，撈出後用冷水衝淨；紅棗略沖洗乾淨；薑略拍。
2. 花菇先泡漲開，用太白粉抓洗片刻，再沖洗乾淨，切片。泡花菇的水保留備用。
3. 湯鍋中放入水7杯（包括香菇水）、雞爪、雞胸、花菇、紅棗和薑，淋下酒，煮滾後蓋上鍋蓋，轉小火煮約40-50分鐘。
4. 加入山藥，再煮約8-10分鐘，熄火。
5. 加入鹽調味即可。

Ingredients

400g Chinese yam
10 chicken feet
1 chicken breast (with bones)
6 Chinese mushrooms
8-10 red dates
1 piece ginger
some potato starch

Seasonings

1 tbsp wine
dash of salt

Method

1. Peel Chinese yam and cut into pieces. Scald chicken feet and breast meat in boiling water for about 1 minute. Remove and rinse under cold water. Briefly rinse red dates. Lightly smash ginger.
2. Rehydrate mushrooms. Wash and rub with potato starch, then rinse clean and cut into slices. Reserve mushrooms soaking water.
3. Pour 7 cups of water into a soup pot (including mushrooms soaking water). Add chicken feet, breast meat, mushrooms, red dates and ginger. Pour in wine and bring to a boil. Put the lid on and simmer over low heat for about 40-50 minutes.
4. Add Chinese yam and cook for another 8-10 minutes. Turn off the heat.
5. Season with salt to taste.

白蘿蔔
Radish

蘿蔔糕
Carrot Cake

白蘿蔔牛腩煲
Braised Radish and Beef Brisket

炒蘿蔔乾
Fried Dried Radish

主要保健功效：
Health Benefits:

含豐富的芥子油、維生素A、C、蛋白質、醣類、鈣、鐵等。具有美白、增進食慾、清熱、止渴解膩、預防高血壓和冠心病等功效。

Rich in mustard oil, vitamins A, C, protein, sugars, calcium and iron. Helps in whitening the skin, increases appetite, dispels heat, quenches thirst, refreshes your palate, prevents high blood pressure and coronary heart disease.

營養烹調祕訣：
Nutritional Facts and Cooking Tips:

日本料理中的白蘿蔔泥、韓國和中國人的白蘿蔔泡菜，都不陌生吧！這樣的吃法，可以不破壞其中活性營養成分。其營養成分很容易在室溫中流失，所以不宜久置。另外，白蘿蔔的葉子含有豐富的鐵質，有助預防貧血和癌症。若進食補品或藥物，應盡量避免吃白蘿蔔，以免減低補益效果。

The nutrients in radish deteriorate easily in room temperature and do not keep well. The pickled radish commonly served in Japanese restaurants and the kim chi in Korean and Chinese restaurants are both clever ways of preserving the nutritional values of the radish. The leaves of the radish are rich in iron, which is good for preventing anemia and cancer. Do not eat radish when you are taking tonics or medicine as it might affect the effectiveness of the medicine.

白蘿蔔還可以這樣吃！
Serving Suggestions:

白蘿蔔適合燉肉、紅燒、煮湯、醃製成泡菜、做成糕點或打汁。熟悉的菜餚有白蘿蔔排骨湯、白蘿蔔燉羊肉、紅燒白蘿蔔、蘿蔔糕等等。

Suitable for stewing with meat, braising, cooking in soups, made into kim chi, made into snacks or for juicing. Common dishes include radish and pork ribs soup, radish stew with lamb, braised radish and carrot cake.

白蘿蔔燜
Braised Radish and Fish Paste

材料

白蘿蔔 250克
魚膠 180克
蒜泥、薑絲 各20克
太白粉水 少許

調味料

高湯 1 1/2杯
蠔油、酒 各1/2大匙
鹽、胡椒粉、麻油適量

做法

1. 白蘿蔔去皮切粗絲，燙煮約10分鐘；魚膠用少許油煎黃兩面，取出後切小塊。
2. 以1大匙油炒香蒜泥和薑絲，加入白蘿蔔、魚膠和調味料，蓋上鍋蓋，燜煮約2分鐘，加入太白粉水勾芡即可。

Ingredients

250g radish
180g minced fish paste
20g each minced garlic, ginger shreds
some potato starch mixture

Seasonings

1½ cups stock
½ tbsp each oyster sauce, wine
dash of salt, pepper, sesame oil

Method

1. Peel radish and cut into thick shreds. Blanch for about 10 minutes. Pan-fry minced fish paste in a little oil until golden brown on both sides, then cut into small pieces.
2. Heat 1 tbsp of oil and fry minced garlic and ginger until fragrant. Add radish, fish paste and Seasonings. Cover with a lid and cook for about 2 minutes. Lastly thicken with potato starch mixture.

蘆筍
Asparagus

蘆筍肉卷湯
Asparagus Meat Rolls Soup

蘆筍扁魚酥
Asparagus Crispy Fish

芥末拌蘆筍
Asparagus with Mustard Dressing

雜錦炒脆蔬
Stir-fried Mixed Vegetables

主要保健功效：
Health Benefits:

含豐富葉酸、維生素B群、C、E、胡蘿蔔素、鐵、鈣等。蘆筍的營養成分有護眼和改善心血管疾病等功效；其豐富的葉酸更有助胎兒健康成長，是孕婦重要的葉酸來源。

An excellent source of folate, B-group vitamins, vitamins C, E, beta-carotene, iron and calcium. It has nutrients good for vision and for protecting against cardiovascular disease. It is especially high in folate which is essential to the healthy development of babies in pregnancy.

營養烹調秘訣：
Nutritional Facts and Cooking Tips:

蘆筍不宜生吃，它是鮮嫩的蔬菜，所以不宜久煮久放，否則營養、甜味俱減低、流失！蘆筍的葉酸很容易被破壞，應盡量避免高溫烹調。蘆筍中普林的含量高，如有痛風，泌尿道結石者不建議食用。

Asparagus should not be eaten raw. It is a delicate vegetable which needs only be cooked lightly. Overcooking or storing them for too long will decrease their nutrients and sweetness. Refrain from cooking under high heat to prevent loss of folate. Asparagus has a high purine content, thus they are not suitable for patients suffering from gout or urinary stones.

蘆筍還可以這樣吃！
Serving Suggestions:

適合清炒或加配料一起炒、汆燙後做沙拉、蔬菜捲。熟悉的菜餚有清炒蘆筍、參巴蘆筍、蘆筍炒蝦仁、肉捲蘆筍等等。

Suitable for stir-frying on its own or with other ingredients. May also be blanched before making into salads or vegetable rolls. Common favourites include stir-fried asparagus, sambal asparagus, stir-fried asparagus with prawns, asparagus meat rolls.

蘆筍炒蝦仁
Stir-fried Asparagus and Prawns

材料

蘆筍10支、蝦仁200克、蘑菇4朵、紅蘿蔔片少許、蔥段少許、薑3片、太白粉水適量

調味料

鹽1茶匙、雞粉1/2茶匙、酒1/2大匙

做法

1. 蘆筍去除老皮，切成2段，放入滾水中燙熟，水中加少許鹽，撈出後衝冷水，瀝乾；蘑菇切半。
2. 燒熱少許油，放入蔥和薑炒香，加入蘆筍、蝦仁、蘑菇、紅蘿蔔拌炒片刻，加入少許水快炒，再加入調味料調味，最後加入太白粉水勾芡即可。

Ingredients

10 asparagus, 200g prawns, 4 button mushrooms, some carrot slices, some spring onion sections, 3 slices ginger, some potato starch mixture

Seasonings

1 tsp salt, ½ tsp chicken seasoning powder, ½ tbsp wine

Method

1. Peel off tough skins from the asparagus and cut each into 2 sections. Blanch in lightly salted boiling water, then rinse under cold water, drain well. Cut mushrooms into halves.
2. Heat some oil and fry spring onion and ginger until fragrant. Stir in asparagus, prawns, mushrooms and carrot. Add a little water and stir briskly. Add Seasonings and lastly thicken with potato starch mixture.

馬鈴薯 Potato

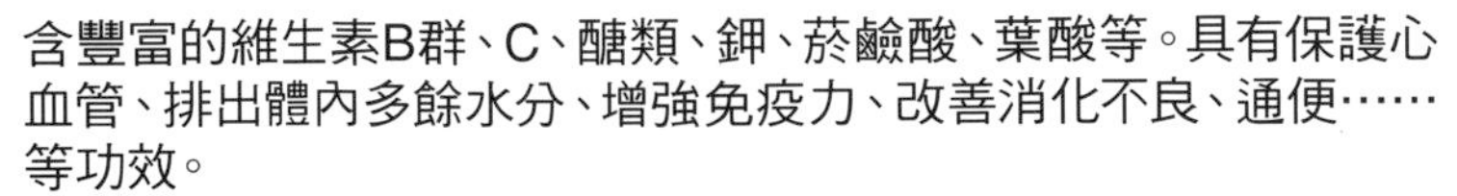

主要保健功效：

Health Benefits:

含豐富的維生素B群、C、醣類、鉀、菸鹼酸、葉酸等。具有保護心血管、排出體內多餘水分、增強免疫力、改善消化不良、通便……等功效。

High in B-group vitamins, vitamin C, sugars, potassium, niacin and folate. Beneficial for good cardiovascular health, helps expel excess fluid from the body, boosts immunity, aids digestion and prevents constipation.

營養烹調祕訣：

Nutritional Facts and Cooking Tips:

千萬不要選購或食用已經發芽或表皮轉綠的馬鈴薯！馬鈴薯中含有極少有毒成分的龍葵素，少含量對人體不會造成傷害，但發芽或轉綠馬鈴薯的龍葵素比平常高出很多倍，食用後恐怕會造成身體不適。

Potato contains a toxin called solanine, and in a much higher concentration in green potatoes and new sprouts. Never eat potatoes that have sprouted or have turned green below the skin as these will make you sick.

絞肉馬鈴薯
Potato with Minced Pork

材料

馬鈴薯	450克（去皮，切絲）
絞肉	150克
蔥花	2大匙

調味料

醬油	1大匙
鹽	適量

做法

1. 鍋中放3大匙油燒熱，放入絞肉炒至轉色，加入蔥花同炒，淋下醬油炒勻。
2. 加入馬鈴薯和水（水要蓋過馬鈴薯），煮至滾後轉小火，蓋上鍋蓋，煮約15分鐘至馬鈴薯熟軟，加鹽調味即可。

Ingredients

450g potato (peeled and shredded)
150g minced pork
2 tbsps chopped spring onion

Seasonings

1 tbsp dark soy sauce
dash of salt

Method

1. Heat 3 tbsps of oil, fry minced pork until it changes colour, toss in spring onion and drizzle over dark soy sauce.
2. Add potato and water (the water level has to be above the potato), bring it to a boil. Turn to low heat and put the lid on. Cook for about 15 minutes until potato is soft. Add salt and serve.

馬鈴薯洋蔥湯
Potato Onion Soup

材料

排骨	250克
馬鈴薯	1個
番茄	1個
紅蘿蔔	1支
洋蔥	1/2個
胡椒粒	1/2大匙
鹽	適量

做法

1. 馬鈴薯、紅蘿蔔、洋蔥、番茄切塊；排骨汆燙去血水。
2. 將所有材料放入湯鍋中，加入水6杯，以大火煮滾後，轉小火煲約1小時，加入鹽調味即可。

Ingredients

250g pork ribs
1 potato
1 tomato
1 carrot
½ onion
½ tbsp peppercorns
dash of salt

Method

1. Cut potato, carrot, onion and tomato into pieces. Scald pork ribs to get rid of blood.
2. Place all ingredients in a soup pot and add 6 cups of water. Bring to a boil over high heat, then turn down the heat and simmer for about 1 hour. Season with salt to taste.

黃瓜 Cucumber

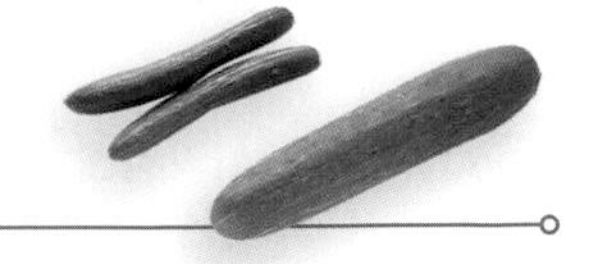

主要保健功效

Health Benefits:

常見的黃瓜有大黃瓜、小黃瓜和老黃瓜，它們的營養成分非常相似，維生素C和含水量高，具有消暑解渴、抗菌消炎、生津潤喉、美白淡斑、柔膚、利尿、消除水腫……等功效。

These include cucumber, Japanese cucumber and old cucumber, all with similar nutritional values. It has a high vitamin C and water content good for relieving summer heat and quenching thirst. It has anti-bacteria properties and helps reduce inflammation, increases bodily fluids, moistens the throat, whitens and softens skin, lightens dark spots, promotes urination and relieves water retention.

營養烹調祕訣

Nutritional Facts and Cooking Tips：

小黃瓜含有破壞維生素C的酵素，加醋或適當加熱可將該酵素破壞，以增加維生素C的攝取。黃瓜籽含有豐富抗氧化劑，也是解毒劑，下一次烹調時記得不要丟棄！

Cucumber contains a certain enzyme that destroys vitamin C, adding vinegar or heat up briefly to aid the absorption of vitamin C. Its seeds are rich in anti-oxidants and have toxins removing properties, thus remember not to remove the seeds when you eat cucumber.

冬瓜 Winter Melon

主要保健功效

Health Benefits:

含豐富的膳食纖維、維生素和營養素等，熱量極低，適合減肥者食用。豐富的維生素成分，可以抑制病毒和細胞的活性，有效預防感冒和抗老化。

Rich in dietary fibre, vitamins and other nutrients. Suitable for people watching their weight as it has very low calories. Its high vitamin content helps fight against bacteria and diseases, and is excellent for preventing colds and flu. It also has anti-aging properties.

營養烹調祕訣

Nutritional Facts and Cooking Tips：

冬瓜具有清熱解毒、利尿的功效，如果連皮和籽一起煮更理想。冬瓜肉較涼，但冬瓜皮性溫且富含維生素C，冬瓜籽則有利尿的作用，所以煮冬瓜湯時，最理想就是冬瓜連皮和籽一起煮。

Cooking winter melon together with the skin and seeds increases its dispelling heat, detoxifying and diuretic properties. The flesh is cold in nature, but the skin is warm and contains a good amount of vitamin C while the seeds have diuretic properties. Thus it is advised to cook the winter melon whole when making soups.

絲瓜 Loofah

主要保健功效

Health Benefits:

具有改善便秘、抗老化、美容養顏、健腦、抗壞血病、抗過敏等功效。

Helps treat constipation, strengthens the mind, prevents scurvy and allergies. It is also anti-aging and is good for the complexion.

營養烹調祕訣

Nutritional Facts and Cooking Tips：

絲瓜不宜生吃，而且要煮熟，以免引起身體不適。烹調絲瓜最好現切現做，因為絲瓜水分豐富，切開後營養成分容易隨瓜汁流失。

Not to be eaten raw. Cook it thoroughly before eating. Cut loofah right before cooking so that the juices are not lost to prevent losing most of its nutrients.

蛤蜊絲瓜粥
Clams and Loofah Porridge

材料

絲瓜	300克
蛤蜊	150克
白米	2/3杯
薑末	少許
鹽	1茶匙
高湯	2杯
酒	少許

做法

1. 絲瓜去皮，切薄片；蛤蜊泡入淡鹽水吐沙約1小時。
2. 白米洗淨，加入水5杯，煮滾後，轉小火煮約20分鐘成白粥，加入絲瓜、蛤蜊、薑末和調味料煮沸，繼續煮約5分鐘即可。

Ingredients

300g loofah
150g clams
⅔ cup uncooked rice
some minced ginger
1 tsp salt
2 cups stock
dash of wine

Method

1. Peel loofah and cut into thin slices. Purge clams in lightly salted water for about 1 hour to get rid of sand.
2 Rinse rice and add 5 cups of water. Bring to a boil, then turn down the heat to cook for about 20 minutes into porridge. Add loofah, clams, ginger and Seasonings. Bring to a boil and cook further for about 5 minutes.

絲瓜色澤翠綠的祕訣！

How do you cook loofah so that it stays green and does not turn mushy?

絲瓜盡量不要刨得太深，這樣絲瓜煮好後，不會軟綿綿，色澤也較漂亮。煮絲瓜時不宜用鐵鍋，否則煮好的絲瓜很快變黑，口感跟色澤都不好！此外烹調絲瓜的時間也不宜太久，起鍋前再用少許的鹽調味，也可以避免絲瓜變黑。

When peeling loofah, peel thinly so that it holds its shape better when cooked. Do not use a metal pot to cook loofah or else it will turn black. Do not cook too long and season with a little salt right before finish cooking so that the loofah retains its colour better.

雙耳煮絲瓜
Loofah with Black and White Fungus

Tasty Tips

- 絲瓜以瓜身完整無損傷、顏色淺綠色、瓜紋分明的為佳。
- 購買白木耳時要選微黃的為佳。純白色的白木耳很可能經人工漂白過，最好別食用。
- Choose a loofah that is not bruised on the surface, is light green in colour and has prominent stripes.
- Choose white fungus that is pale yellow in colour as those that are pure white are mostly bleached and should not be used.

材料

絲瓜	500克
乾黑木耳	75克
乾白木耳	75克
紅蘿蔔片	少許
薑片	少許
蔥段	少許
高湯	1杯

調味料

鹽	少許
太白粉水	適量
麻油	1/2茶匙

做法

1. 白、黑木耳分別泡軟，切小片。
2. 絲瓜刨去外皮，不要刨太深，盡量保留綠色部分，切斜片。
3. 鍋中放少許油燒熱，放入蔥段和薑片爆香，加入黑、白木耳和絲瓜拌炒片刻，加入高湯和調味料燜煮約3分鐘。
4. 加入太白粉水勾芡，再加入少許鹽調味，最後滴入麻油即可。

Ingredients

500g loofah
75g dried black fungus
75g dried white fungus
some carrot slices
some ginger slices
some spring onion sections
1 cup stock

Seasonings

dash of salt
some potato starch mixture
½ tsp sesame oil

Method

1. Soak black and white fungus until soft and trim into smaller pieces.
2. Peel loofah thinly, keeping the green part as much as possible, then slant cut into slices.
3. Heat oil and fry spring onion and ginger until fragrant. Stir in black and white fungus and loofah to fry briefly. Add stock and Seasonings, simmer for about 3 minutes.
4. Thicken with potato starch mixture, then season with salt to taste. Lastly add sesame oil.

還有哪些食材適合用來釀肉？
What other kinds of vegetables may be used for this dish?

也可以用苦瓜、黃瓜、白蘿蔔、冬瓜、南瓜等食材來釀肉，唯蒸的時間長短不同，需要加以調整。

Bitter gourd, cucumber, radish, winter melon and pumpkin may also be used. However these vegetables take different times to cook, thus adjust the steaming time accordingly.

釀節瓜
Stuffed Hairly Gourd

Tasty Tips

- 在節瓜內圈撒少許太白粉可幫助肉料固定，避免掉落。
- 肉料的配料可隨意變換，如加入馬蹄、三色蔬菜等等。
- Dusting potato starch on the insides of the hairly gourd helps secure the stuffing in place.
- You may vary the ingredients for the stuffing. Water chestnuts and mixed vegetables may be used too.

材料

節瓜	1條
豬絞肉	250克
乾香菇	2朵
紅蘿蔔	1/2支
蔥花	1大匙
蝦米	2大匙
高湯或水	1杯
太白粉水	少許

調味料

蛋	1個
醬油	1大匙
酒	1/2大匙
鹽	1/4茶匙
胡椒粉	少許
太白粉	1/2大匙
水	1大匙

做法

1. 香菇泡軟，去蒂、切丁；紅蘿蔔切丁，燙熟；蝦米泡軟，摘去硬殼。
2. 節瓜削皮後切成2公分寬的圓段，挖除瓜籽（保留少許底部），全部用滾水汆燙約10秒鐘（水中加少許鹽），撈起瀝乾。
3. 絞肉加入香菇、蝦米、紅蘿蔔和蔥花，並加入打散蛋液和調味料，仔細攪拌均勻至有黏性。
4. 在節瓜內圈撒少許太白粉，將肉料填入，抹光表面，放入蒸碗中，加入高湯或水，再加少許鹽和醬油調味。
5. 燒開蒸鍋中的水，放入釀好的節瓜，以大火蒸約25-30分鐘至熟，取出將蒸汁倒入鍋中，稍煮後加入太白粉水勾芡，淋在釀節瓜上即可。

Ingredients

1 hairly gourd
250g minced pork
2 dried mushrooms
½ carrot
1 tbsp chopped spring onion
2 tbsps dried shrimps
1 cup stock or water
some potato starch mixture

Seasonings

1 egg
1 tbsp dark soy sauce
½ tbsp wine
¼ tsp salt
dash of pepper
½ tbsp potato starch
1 tbsp water

Method

1. Soak mushrooms until soft and remove the stems, cut into dices. Cut carrot into dices and blanch in boiling water. Soak dried shrimps until soft, then pluck away the hard bits.
2. Peel hairly gourd and cut into 2cm thick round pieces. Dig out the seeds but do not dig through. Blanch in lightly salted boiling water for about 10 seconds. Remove and drain.
3. Combine well minced pork, mushrooms, dried shrimps, carrot and chopped spring onion. Add beaten egg and Seasonings, stir well until mixture is sticky.
4. Dust a little potato starch on the insides of the hairly gourd and stuff it with the meat mixture. Coat it smooth and place in a steaming dish. Add stock or water, then season with a dash of salt and dark soy sauce.
5. Bring water in the steamer to a boil, steam stuffed hairly gourd over high heat for about 25-30 minutes until cooked through. Remove and drain out the steaming liquid into a pot, cook the liquid briefly and thicken with potato starch mixture. Drizzle over the stuffed hairly gourd to serve.

苦瓜 Bitter Gourd

主要保健功效：

Health Benefits:

含豐富的維生素C、蛋白質、脂肪、澱粉、鈣、磷、鐵、胡蘿蔔素、核黃素等。具有廣為人知的消暑降火、清肝明目、消炎解毒和降血壓等功效。

Rich in vitamin C, protein, fats, starch, calcium, phosphorus, iron, beta-carotene and riboflavin. Known for its cooling effect on the body, bitter gourd also cleanses the liver, brightens the eyes, reduces inflammation, eliminates toxins and lowers blood pressure.

營養烹調祕訣：

Nutritional Facts and Cooking Tips:

苦瓜含水溶性維生素，長時間烹調維生素會流入湯汁裡，適合涼拌、打汁、快炒或煮湯。其維生素C加熱後容易流失，因此不宜煮太久。

Cooking in liquid over a long period of time can result in a huge loss of water-soluble vitamins in bitter gourds. It is suitable for used in salads, juicing, quickly frying or cooking in soups. Do not overcook to prevent loss of vitamin C.

涼拌五味苦瓜
Cold Bitter Gourd with Five-flavours Sauce

材料

苦瓜	1/2個
蔥	1支
蒜	2瓣
紅辣椒	1支
香菜	少許

調味料

醬油	1大匙
番茄醬	1大匙
醋	2大匙
糖	2大匙
鹽	少許
麻油	2大匙

做法

1. 苦瓜去籽和內膜，切薄片，泡入水中，放入冰箱冰2小時。
2. 蔥切碎；蒜磨成泥；紅辣椒去籽、切碎；香菜切細末。
3. 取出苦瓜吸乾水，加入混勻調味料拌勻即可。

Ingredients

½ bitter gourd
1 spring onion
2 clove garlic
1 chilli
some coriander

Seasonings

1 tbsp dark soy sauce
1 tbsp tomato sauce
2 tbsps vinegar
2 tbsps sugar
dash of salt
2 tbsps sesame oil

Method

1. Remove the seeds and pulp of bitter gourd, then slice thinly. Soak in water and keep in the refrigerator for 2 hours.
2. Chop spring onion. Grind garlic into paste. Deseed and chop the chilli. Chop coriander.
3. Remove the bitter gourd from refrigerator and pat dry, mix well with Seasonings mixture.

江魚仔參巴醬燜苦瓜
Bitter Gourd with Ikan Bilis Sambal

材料

苦瓜	1條
蒜	3瓣

調味料

江魚仔參巴醬	1大匙
蠔油	1大匙
糖	1/2茶匙

做法

1. 苦瓜片開兩半，去籽和內膜，切長條狀，燙後用冷水沖涼；蒜切片。
2. 燒熱少許油，以小火爆香蒜片和江魚仔參巴醬，放入苦瓜拌炒片刻，最後加入水50毫升和剩餘調味料，煮滾後轉小火，燜煮約2分鐘，轉大火煮至湯汁略收乾即可。

Ingredients

1 bitter gourd
3 cloves garlic

Seasonings

1 tbsp ikan bilis sambal paste
1 tbsp oyster sauce
½ tsp sugar

Method

1. Cut bitter gourd lengthwise into halves. Remove the seeds and pulp, then cut into long sections. Blanch in boiling water, then rinse under cold water. Cut garlic into slices.
2. Heat oil and fry garlic and sambal paste over low heat until fragrant. Add bitter gourd and stir-fry briefly. Add 50ml water and remaining Seasonings, bring to a boil, then simmer over low heat for about 2 minutes. Turn up the heat and boil over high heat until sauce is reduced.

南瓜
Pumpkin

南瓜濃湯
Pumpkin Soup

南瓜炒米粉
Pumpkin Fried Bee Hoon

南瓜果醬
Pumpkin Jam

主要保健功效：
Health Benefits:

南瓜是優質鹼性蔬菜，從裡到外都是寶，被視為瓜果中的金子。含豐富的維生素A、胡蘿蔔素、鈣、鉀、鉻等礦物質，可預防眼睛疾病和感冒，還有抗氧化、刺激胰島素分泌、降低血糖和血脂等功效。

Pumpkin is the gold among the gourds as it is an excellent alkaline vegetable packed full of goodness. It is rich in vitamin A, beta-carotene, calcium, potassium and chromium. Protects against eye diseases and flu. It is a good anti-oxidant, affects insulin secretion, lowers blood sugar level and reduces blood lipids.

營養烹調祕訣：
Nutritional Facts and Cooking Tips:

南瓜肉帶甜味，烹煮時要注意糖的用量。其外皮和囊含豐富的營養素，最好連這些部分一起食用。烹調時，可加油脂一起烹煮，其類胡蘿蔔素更容易被吸收。

Add sugar sparingly when cooking with pumpkin as its flesh tastes sweet. Do not discard its skin and pulp as they contain valuable nutrients. Cook pumpkin with fats as it increases the absorption of carotenoids.

南瓜還可以這樣吃！
Serving Suggestions:

南瓜耐高溫烹煮，營養素不易流失，所以適合烹煮的方式也很多，無論燉、燜、燒、烤、炸、蒸、煮都合適。熟悉的菜餚有南瓜炒米粉、南瓜濃湯、南瓜粥、南瓜餅等等。南瓜泥也常用來做甜點的餡料。

Pumpkin cooks well under high heat and its nutrients are not easily lost during cooking. Therefore pumpkin is suitable for many different kinds of cooking methods including stewing, braising, grilling, deep-frying, steaming and boiling. Familiar favourites include pumpkin fried bee hoon, pumpkin soup, pumpkin porridge, pumpkin cake. Pumpkin puree is also often used as a filling for sweet desserts.

蝦米南瓜
Pumpkin with Dried Shrimps

材料
南瓜350克、蝦米10粒、薑末1大匙、蔥末1大匙、蒜末1大匙、辣椒乾末1大匙、太白粉水適量

調味料
醬油2大匙、醋1大匙、酒1大匙、糖1大匙

做法
1. 南瓜洗淨，連皮切小塊，放入熱油炸熟，表面呈金黃，撈起瀝乾油；蝦米泡軟，摘去硬殼。
2. 鍋中放少許油燒熱，爆香辣椒乾、薑、蒜、蔥和蝦米，加入調味料，煮至微濃稠，加入南瓜拌炒均勻，最後加入太白粉水勾芡即可。

Ingredients
350g pumpkin, 10 dried shrimps
1 tbsp each chopped ginger, spring onion, garlic, dried chillies, some potato starch mixture

Seasonings
2 tbsps dark soy sauce, 1 tbsp vinegar, 1 tbsp wine, 1 tbsp sugar

Method
1. Rinse pumpkin and cut into small pieces, leaving the skin on. Deep-fry in hot oil until cooked and golden brown. Remove and drain away the oil. Soak dried shrimps until soft, then pluck away the hard bits.
2. Heat oil and fry dried chillies, ginger, garlic, spring onion and dried shrimps until fragrant, add Seasonings. When the sauce has slightly thickened, add pumpkin and stir well. Lastly thicken with potato starch mixture.

茹菌
Fungi

烤什錦菇
Grilled Mushrooms

什錦菇湯
Mixed Mushrooms Soup

雙菇燒麵筋
Braised Mushrooms and Wheat Gluten

主要保健功效：
Health Benefits:

茹菌類的主要營養成分包括維生素B群、D、蛋白質、糖類、鋅、鈣等。具有改善便秘、降血壓和膽固醇、增強免疫能力、促進成長發育、預防骨骼疏鬆等功效。

The main nutrients include B-group vitamins, vitamin D, protein, sugars, zinc and calcium. Helps treat constipation, lowers blood pressure and cholesterol, improves body's resistance to diseases, promotes growth and prevents osteoporosis.

營養烹調祕訣：
Nutritional Facts and Cooking Tips:

茹菌類不宜生吃，要煮熟才能食用。且每次食用的量不可太多，因為茹菌類含有高纖維，吃多了可能會導致腹瀉。

It is advised to cook mushrooms before consuming. Do not eat in excess as mushrooms have a very high fibre content and might cause diarrhea.

茹菌類還可以這樣吃！
Serving Suggestions:

茹菌類品種繁多，常見的有香菇、草菇、金針菇、杏鮑菇、鴻禧菇、蘑菇等。

可加入蔬菜拌炒、煮湯、做火鍋、燜煮。熟悉的菜餚有釀香菇、木耳雞、素炒菇、鮮菇雞湯、茹菌天婦羅等等。

There are many different types of mushrooms in the fungi family, common ones include the shiitake mushrooms, straw mushrooms, enoki mushrooms, king oyster mushrooms, hon-shimeji mushrooms and button mushrooms.

Suitable for stir-frying with vegetables, cooking in soups or steamboat and braising. Common favourites include stuffed mushrooms, chicken and fungus, vegetarian fried mushrooms, chicken and mushrooms soup, mushrooms tempura.

什錦炒菇菌
Stir-fried Assorted Mushrooms

材料
新鮮香菇3朵、杏鮑菇1朵、鴻禧菇1大朵、金針菇1把、巴西蘑菇5朵、蔥段少許、蒜末2大匙

調味料
蠔油2大匙、水2大匙、麻油1茶匙

做法
1. 香菇和杏鮑菇切片；鴻禧菇和金針菇切除根部，分成小朵；巴西蘑菇切半。
2. 燒熱少許油，爆香蔥段和蒜末，放入所有菇菌、蠔油和水，炒至菇菌熟透，最後滴入麻油即可。

Ingredients
3 fresh mushrooms, 1 king oyster mushroom, 1 bunch hon-shimeji mushrooms, 1 bundle enoki mushrooms, 5 agaricus blazei murill mushrooms, some spring onion sections, 2 tbsps minced garlic

Seasonings
2 tbsps oyster sauce, 2 tbsps water, 1 tsp sesame oil

Method
1. Cut fresh mushrooms and king oyster mushroom into slices. Trim away the roots of the hon-shimeji mushrooms and enoki mushrooms, cut into small florets. Cut agaricus blazei murill mushrooms into halves.
2. Heat oil and fry spring onion and garlic until fragrant. Add all the mushrooms, oyster sauce and water, fry until mushrooms cook through. Lastly drizzle over sesame oil.

切牛肉有技巧！

How to cut beef for the best texture?

牛肉切片時宜逆紋切，即與肉紋呈垂直方向切，這樣肉的纖維被切斷，烹調時就不會緊縮變小。

Slice beef across the grain, that is, cut perpendicular to the muscle fibers so that the meat will not contract and become tough when cooked.

金針菇牛肉捲
Beef Rolls with Enoki Mushrooms

Tasty Tips

- 也可將金針菇換成蘆筍、鴻禧菇等。
- 牛肉捲也可用來烤，變換另一新菜色。
- *Asparagus and hon-shimeji mushrooms may be used in place of enoki mushrooms.*
- *You may also grill the beef rolls to make a different dish.*

材料

牛肉	200克
洋蔥	30克
金針菇	100克
菜心	120克

調味料

鹽	少許
糖	少許
日本照燒醬汁	2大匙

做法

1. 牛肉切成薄片；洋蔥去皮，切成細條；金針菇去除根部，稍微沖洗，瀝乾水。
2. 燒熱少許油，放入洋蔥和金針菇稍拌炒，加入鹽和糖拌勻，撈出。
3. 牛肉片攤平，放入金針菇和洋蔥，捲起。
4. 燒熱少許油，放入牛肉捲煎至香且金黃色，撈出。
5. 菜心放入沸水中燙熟，撈出盛盤，放上牛肉捲。
6. 淋上照燒醬汁即可。

Ingredients

200g beef
30g onions
100g enoki mushrooms
120g choy sum

Seasonings

dash of salt
dash of sugar
2 tbsps teriyaki sauce

Method

1. Cut beef into thin slices. Remove skin and cut onions into thin strips. Trim away the roots of enoki mushrooms. Rinse and drain.
2. Heat some oil, stir-fry onions and enoki mushrooms briefly, add salt and sugar. Remove.
3. Place a slice of beef flat, place onions and enoki mushrooms on the meat, roll up.
4. Heat some oil, pan-fry beef rolls until fragrant and golden in colour. Remove.
5. Blanch choy sum in boiling water, remove. Transfer to the plate and top with beef rolls.
6. Drizzle teriyaki sauce on top and serve.

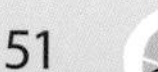

四季豆 French Bean

主要保健功效
Health Benefits:

可促進造血功能，並改善貧血症狀。其非水溶性的膳食纖維可促進腸胃蠕動，幫助排便順暢。四季豆是一種難得的高鉀、高鎂、低鈉的蔬菜，尤其適合心臟病、動脈硬化、高血壓患者食用。

Aids in blood formation and improves anemia. The insoluble fibre helps in bowel movements and prevents constipation. French bean is high in potassium and magnesium but low in sodium, thus it is especially suitable for people with heart disease, atherosclerosis and hypertension.

營養烹調祕訣
Nutritional Facts and Cooking Tips：

適合炒、煸和煮，四季豆要煮熟透才食用，因為其中含有兩種成分，若沒有熟透可能會引起噁心、嘔吐或腹痛等症狀。

Suitable for stir-frying, dry-frying and boiling. Cook French beans thoroughly before eating, eating them raw or half-cooked might cause nausea or stomach upset.

長豆 Long Bean

主要保健功效
Health Benefits:

含豐富的蛋白質、脂肪、膳食纖維、維生素B、C等。可促進消化腺和胰島素的分泌，也可以促進腸胃蠕動，增進食慾。

An excellent source of protein, fats, dietary fibre, vitamins B, C. Regulates the function of digestive gland and secretion of insulin. Promotes normal bowel movement and increases appetite.

營養烹調祕訣
Nutritional Facts and Cooking Tips：

適合炒、煮湯、煮粥、汆燙後涼拌等，不適合長時間烹調，以免營養流失。

Good for stir-frying, cooking in soups and porridge, blanching then used in salads. Do not overcook to prevent loss of nutrients.

豌豆 Pea

主要保健功效
Health Benefits:

有抗菌消炎、增強新陳代謝、養脾、養陰、止渴等功效。脾胃較弱者不適合多吃，以免脹氣。

Fights bacteria and stops infection, enhances metabolism, nourishes spleen and yin, helps quench thirst. Do not take in excess if you have a weak stomach as it causes gas.

營養烹調祕訣
Nutritional Facts and Cooking Tips：

豌豆家族包括：豌豆莢、甜豆莢、豌豆仁、豌豆苗、豌豆嬰。適合炒、煮、做成配料，增添色彩。和所有豆類一樣，務必煮熟煮透才可食用，以去除有害物質。和富含氨基酸的食材一起烹煮，可以攝取豌豆的營養價值。

The pea family includes pea pods, sweet peas, peas, pea shoots and young pea shoots. Suitable for stir-frying, boiling and as a side dish. Cook through before eating. Cook with foods high in amino acids to attain the full nutrients.

秋葵 Lady's Finger

主要保健功效

Health Benefits:

其獨特的黏性液汁和豐富的營養素對消化系統有益，具有健胃整腸、保護肝臟、美白肌膚、抗老化的功效。也有蔬菜中的「威爾剛」之稱。

Its sticky substance and rich nutrients have beneficial effects on our digestive systems. It strengthens the stomach, protects the liver, beautifies complexion and has anti-aging properties. Also known for improving erectile dysfunction condition.

營養烹調祕訣

Nutritional Facts and Cooking Tips：

適合炒、燴、煮湯、汆燙後做沙拉等。秋葵營養豐富，也很嬌貴，儲存時溫度不宜太高，否則容易腐爛；碰撞後很快變黑，買回來後最好盡快烹調食用。

Suitable for stir-frying, braising, cooking in soups, blanching then used in salads. Lady's finger does not keep well. It deteriorates quickly in high room temperature and bruises and turns black easily.

辣蝦米秋葵

Stir-fried Lady's Finger with Spicy Dried Shrimp

材料A

秋葵 10條
巴拉煎 30克（烘香、壓碎）
鹽 適量
糖 適量

材料B

蝦米 50克（泡軟）
紅辣椒 50克
小辣椒 20克
小紅蔥 25克
蒜 25克
月桂豆 2粒

做法

1. 秋葵放入滾水中燙熟，撈出盛盤；材料B放入攪拌器攪成泥。
2. 鍋中放油燒熱，放入材料B和巴拉煎，以小火不停翻炒至辛香味和油溢出，加入鹽和糖調味，淋在秋葵上即可。

Ingredients A

10 lady's fingers
30g belacan (toasted, mashed)
dash of salt
dash of sugar

Ingredients B

50g dried shrimp (soft till soft)
50g red chilli
20g chilli padi
25g shallots
25g garlic
2 candlenuts

Method

1. Blanch lady's fingers in boiling water until cooked. Remove. Blend Ingredients B into a paste using a food processor.
2. Heat some oil and fry Ingredients B and belacan over low heat, stirring constantly until a pungent smell and oil seeps out. Season with salt and sugar and pour over the lady's fingers.

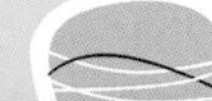

泡粉絲要用冷水或溫水？
Do you soak tung hoon in cold or warm water ?

泡粉絲的水溫可視烹調方式而定，若要再煮入味的，以冷水泡軟為佳，因為粉絲下鍋烹調時，可以再膨漲和吸味。如果不要粉絲下鍋後吸收太多湯汁，可用溫水泡軟，使粉絲多膨漲一點。

Soak tung hoon in cold water if you want it to further soak up the flavours and sauce of the dish that you are cooking it in. If you are not cooking it in a sauce, soak in warm water instead to allow it to swell more before cooking.

四季豆燒粉絲
Braised French Beans with Tung Hoon

Tasty Tips

- 炒四季豆時要用大火且快速拌炒，以免顏色變黃、變黑，用旺火炒也可以避免四季豆表面熟了，裡面仍半生不熟。
- 粉絲泡軟後，下鍋前先剪短一點，比較好夾取。
- *It is important to fry French beans over high heat so that they don't turn yellow or black. This will also ensure that the beans cook through properly.*
- *For easy handling when stir-frying the dish, cut the tung hoon shorter when they have softened.*

材料

四季豆	450克
絞肉	80克
粉絲	2把
蔥末	1大匙

醃料

醬油	1大匙
鹽	1/2茶匙
水	1杯
麻油	少許

做法

1. 四季豆摘除頭尾兩端和莢邊的粗筋，切段；粉絲泡軟。
2. 燒熱3大匙油，放入四季豆，以大火炒約3分鐘，至四季豆變軟，盛出。
3. 放入絞肉炒散，加入蔥末拌炒片刻，淋下醬油、鹽和水，再加入四季豆燒煮約3分鐘。
4. 加入粉絲再燒煮約2-3分鐘，如湯汁仍多，可轉大火收乾湯汁，最後滴下麻油即可。

Ingredients

450g French beans
80g minced pork
2 bundles mung bean vermicelli (tung hoon)
1 tbsp chopped spring onion

Marinade

1 tbsp dark soy sauce
½ tsp salt
1 cup water
dash of sesame oil

Method

1. Trim the two ends and tough fibers of the French beans, cut into sections. Soak tung hoon until soft.
2. Heat 3 tbsps of oil and fry French beans over high heat for about 3 minutes until they soften. Remove.
3. Add minced pork and fry until the meat separates, toss in chopped spring onion, add dark soy sauce, salt and water. Lastly add French beans and cook for about 3 minutes.
4. Add tung hoon and continue to cook for another 2-3 minutes. Turn up the heat to reduce the gravy if you prefer. Lastly add sesame oil.

去除粗筋功夫不可省！

Be sure to trim away the tough fibers.

四季豆在烹調前，除了要清洗乾淨，一定要去除莢邊的粗筋，以免影響口感，也不利消化。

Trim away the stringy tough fibers when preparing French beans for cooking as they will affect the texture of the dish and also are difficult to digest.

四季豆魚麵
Noodles with French Beans and Fish

Tasty Tips

- 若覺得魚去骨麻煩的話，可以整條煮，但是吃的時候要小心魚刺。
- 把煮熟的魚從湯中取出剝下魚肉，最後再放回鍋中煮，可避免魚肉煮太久，並保持魚肉的鮮味。
- To make work easier, cook noodles with the whole fish, but be careful of bones when eating the noodles.
- Remove fish meat from the fish and then return it back to the soup at a later stage helps prevent the fish from overcooking and turning tough.

材料

四季豆	150克
赤棕魚	2條（約300克）
蔥	2支
薑	2片
細麵條	300克

調味料

酒	1大匙
鹽	適量
胡椒粉	少許

做法

1. 魚處理乾淨，擦乾水；四季豆摘除頭尾兩端和莢邊的粗筋，斜切成絲；蔥切段。
2. 燒熱3大匙油，放入蔥段和薑片爆香，再放入魚略煎，淋酒並加入6杯水，煮滾後轉小火煮約5分鐘。
3. 夾出魚，待涼後細心剝下兩面的魚肉（盡量保持大片），再將魚頭、魚骨和肚子部位放回湯中，熬煮約20分鐘至入味，用細篩網過濾湯汁。
4. 細麵煮熟，過一下冷水後，放入魚湯中，再加四季豆和調味料，小火煮約5分鐘，放入魚肉再煮，一滾即可盛碗。

Ingredients

150g French beans
2 red sea bream (about 300g)
2 stalks spring onion
2 slices ginger
300g thin noodles

Seasonings

1 tbsp wine
dash of salt
dash of pepper

Method

1. Rinse fish and pat dry. Trim the two ends and tough fibers of the French beans, then cut into shreds. Cut spring onion into sections.
2. Heat 3 tbsps of oil and fry spring onion and ginger until fragrant. Briefly pan-fry fish, then drizzle wine and add 6 cups of water. Bring to a boil, then cook over low heat for about 5 minutes.
3. Remove fish and gently slice out the fillet on both sides of the fish, refrain from breaking up the fish. Return the fish head, bones and carcass back into the soup. Cook for a further 20 minutes. Strain soup over a fine strainer.
4. Cook noodles and rinse under cold water, place into the fish soup. Add French beans and Seasonings to cook over low heat for about 5 minutes. Return fish to the soup and bring to a boil before serving.

茄子 Eggplant

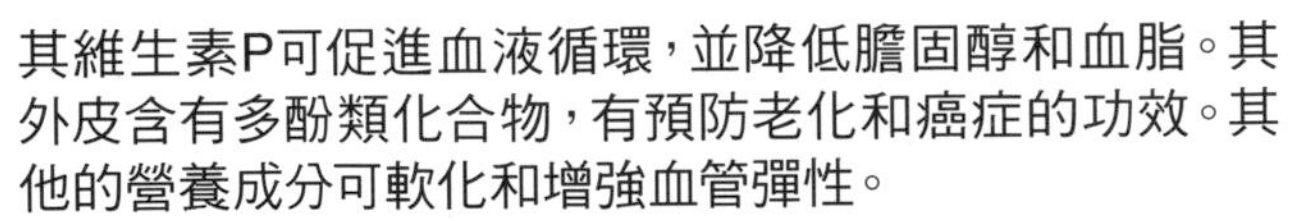

主要保健功效
Health Benefits:

其維生素P可促進血液循環，並降低膽固醇和血脂。其外皮含有多酚類化合物，有預防老化和癌症的功效。其他的營養成分可軟化和增強血管彈性。

Has a high content of vitamin P which helps regulate blood flow, lowers cholesterol and reduces blood lipids. The skin is rich in anti-aging and cancer-fighting flavonoids. It also has properties for softening and maintaining vascular elasticity.

營養烹調祕訣
Nutritional Facts and Cooking Tips：

茄子種類很多，黑紫色茄子是許多營養師推薦的好食材。茄子放久，營養價值也會流失，所以要新鮮食用。茄子皮含有豐富維生素P，烹調時最好連皮一起煮。

Eggplants come in many varieties and the dark purple eggplant is top on the list in terms of nutritional value. Do not keep eggplant for too long as it will losing most of its nutrients. The skin contains a great deal of vitamin P, thus it is important to cook it with the skin on.

甜椒 Capsicum

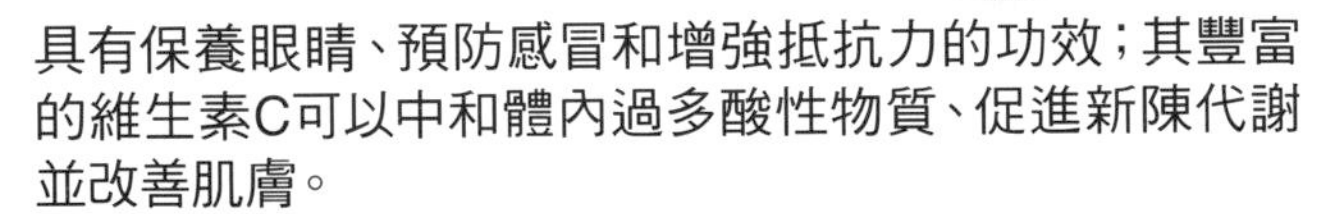

主要保健功效
Health Benefits:

具有保養眼睛、預防感冒和增強抵抗力的功效；其豐富的維生素C可以中和體內過多酸性物質、促進新陳代謝並改善肌膚。

Good for healthy eyes, prevents colds and enhances immunity. Its rich vitamin C content helps restore the pH balance in the body, promotes metabolism and improves complexion.

營養烹調祕訣
Nutritional Facts and Cooking Tips：

甜椒適合生吃或打汁，營養成分不被破壞，也可以更完整攝取。生食要清洗乾淨，除蒂去籽，以免有農藥殘留。如要烹調，可以大火快炒，迅速起鍋。

Suitable for eating raw or juiced to best preserve its full nutritional benefits. When eaten raw, be sure to wash it thoroughly to remove as much produce pesticide residue as possible, then cut away the stem and the seeds. If you prefer to cook it, simply stir-fry briefly over high heat.

番茄 Tomato

主要保健功效
Health Benefits:

具有抗氧化、清除自由基、降低血壓和預防心臟疾病等功效。未成熟的青番茄，不宜生食，因為其含有單寧酸和龍葵素，容易產生腸胃負擔，引起噁心、嘔吐現象，建議至少要烹煮過後再吃較為安全。

It has anti-oxidant properties, eliminates free radicals, lowers blood pressure and prevents heart diseases. Green tomatoes contain tannin and solanine which cause stomach upset and nausea, thus it is best to eat it when it has ripened or cook unripened ones before eating.

營養烹調祕訣
Nutritional Facts and Cooking Tips：

番茄適合生或熟食，生食可攝取較多維生素C，熟食則可攝取較多茄紅素。其維生素A屬脂溶性，最好隨含油脂的食物一起吃，維生素A更容易被吸收。

Tomatoes are delicious both eaten raw or cooked. Eat raw to attain most of the vitamin C, however our body absorbs lycopene more easily if the tomatoes have been cooked. Eat tomatoes with oily foods so that its fat-soluble vitamin A is more easily absorbed by the body.

蔬菜天婦羅
Vegetable Tempura

材料

新鮮香菇4朵、蘆筍4支、茄子1個、番薯1個、甜椒1個、秋葵1個

麵糊

低筋麵粉100克、玉米粉20克（混勻）

冰水	150毫升
蛋	1個
鹽	1/2茶匙

做法

1. 將各種蔬菜切成片或段；蛋和冰水拌勻，加入混勻粉料攪拌均勻，加鹽調味。
2. 蔬菜沾裹麵糊，放入150°C的油中，以小火炸約3分鐘，撈出瀝乾油。可沾點蘿蔔泥醬油食用。

Ingredients

4 fresh mushrooms, 4 asparagus, 1 eggplant,
1 sweet potato, 1 capsicum, 1 lady's finger

Batter

100g cake flour, 20g corn flour (mixed together)
150ml ice water
1 egg
½ tsp salt

Method

1. Cut vegetables into slices or sections. Combine well egg and ice water, then mix well with the combined flour into a batter. Add salt.
2. Coat vegetables with batter and deep-fry in 150ºC oil over low heat for about 3 minutes. Remove and drain well. Serve with soy sauce and radish dipping sauce.

蔬菜就該這樣炒

作　　者　美食編輯企劃小組
發 行 人　程安琪
總 策 劃　程顯灝
總 編 輯　潘秉新
副總編輯　呂增娣
封面設計　陳淑瑩

出 版 者　橘子文化事業有限公司
總 代 理　三友圖書有限公司
地　　址　106台北市安和路2段213號4樓
電　　話　(02) 2377-4155
傳　　真　(02) 2377-4355
E-mail　service@sanyau.com.tw
郵政劃撥　05844889 三友圖書有限公司

總 經 銷　貿騰發賣股份有限公司
地　　址　新北市中和區中正路880號14樓
電　　話　（02）8227-5988
傳　　真　（02）8227-5989

國家圖書館出版品預行編目資料

蔬菜就該這樣炒！／美食編輯企劃小組作. -- 初版. --
臺北市：橘子文化, 2011.01
面；　公分
ISBN 978-986-6890-90-1（平裝）
1.蔬菜食譜 2.烹飪 3.養生

427.3　　99026561

初　　版　2011年1月10日
定　　價　新臺幣149元
ISBN　978-986-6890-90-1